The equiangular sundial at the Royal Greenwich Observatory, Herstmonceux Castle, East Sussex, designed by Gordon E. Taylor. The hour-ring has the hours and minutes marked out in equal divisions. The gnomon is a vertical rod, slotted into the dial axis, which is moved up or down the axis on a date scale to ensure that the shadow indicates the right time.

SUNDIALS

Christopher St J. H. Daniel

Shire Publications Ltd

CONTENTS

Set in 9 point Times roman and printed in Great Britain by C. I. Thomas & Sons (Haverfordwest) Ltd, Press Buildings, Merlins Bridge, Haverfordwest, Dyfed.

British Library Cataloguing in Publication Data available.

ACKNOWLEDGEMENTS

The author wishes to thank friends and colleagues for their help in the preparation of this book. Photographs are acknowledged as follows: the Director, Royal Greenwich Observatory, Herstmonceux, page 1; Cadbury Lamb, page 13 (left); Jack Minnitt, Sun Life Assurance, Bristol, page 14 (left); G. P. Woodford, Pont L'Abbé, France, page 26 (lower right). Other illustrations are photographs taken by the author or are taken from works belonging to the author. The author wishes to thank all those persons or bodies who allowed him to take photographs on their premises, in particular the Trustees and Director of the National Maritime Museum, Greenwich, for permission to reproduce the cover photograph and the photographs (taken by the author) on pages 15 (both) and 17 (left), as well as the engraving (D1132) on page 24 (lower).

COVER: *A modern equinoctial mean-time sundial, the dolphin dial at the National Maritime Museum, Greenwich. The dial plate is engraved with the equation of time correction curves. The gap between the shadows of the dolphin's tails indicates clock time to within a minute. The dial was designed by the author in 1977, to celebrate Her Majesty the Queen's Silver Jubilee, and was sculpted by Edwin Russell (Brookbrae of London).*

BELOW: *The equation of time. Sundials indicate local apparent solar time. To obtain local mean solar time follow the correction curve until it coincides with the date on the horizontal scale. Read off the correction in minutes on the vertical scale, designated plus or minus. Apply this correction to the time shown by the sundial. To obtain Greenwich Mean Time allow a correction for the longitude of the sundial at the rate of four minutes for every one degree of longitude, plus for longitude west, minus for longitude east. When British Summer Time is in force add one hour.*

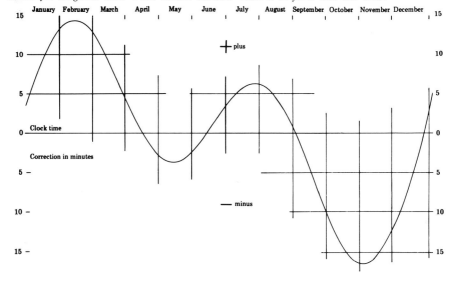

A seventeenth-century common or garden horizontal sundial which has been taken and used as a keyhole plate for the local church at Sydenham, Oxfordshire. (It does not work as a vertical dial.)

INTRODUCTION

Sundial hunting does not require much effort, but it is an addictive pursuit. The purpose of this book is to help people to know what to look for and to understand what they are looking at when they find it.

Sundials fall into two main categories, namely *fixed* sundials and *portable* sundials. Fixed sundials are those which are normally found outside, attached to a building, fixed to a pillar, or on a pedestal in a garden. Portable sundials, though nowadays usually kept indoors, used to be carried around by gentlemen of leisure before the days of pocket watches. Such dials are now mostly collectors' pieces and some can be seen in museums. This book is concerned only with fixed sundials.

A sundial is an instrument which determines the time from the sun by using an indicator, called a *gnomon* or *style*, to cast a shadow or to project a spot of light on to a graduated surface. The time obtained (sundial time) is termed *apparent solar time*. It differs from clock time by an amount known as the *equation of time*, caused partly by the varying speed at which the earth travels in its path round the sun, and due partly to the fact that the axis of the earth is tilted some 23½ degrees to the plane of its orbit. Clock time is a convenient but entirely artificial *mean time* of these variations and for national if not commercial reasons is normally based upon a standard time zone, such as Greenwich, whence the term *Greenwich Mean Time* (GMT). However, when applying the equation of time correction to a sundial reading to obtain clock time, the actual mean time obtained will be *local mean*

3

The Signs of the Zodiac				
Season	Symbol	Latin name	English name	Date of Sun's entrance
SPRING SIGNS	♈	Aries	The Ram	March 21
	♉	Taurus	The Bull	April 20
	♊	Gemini	The Twins	May 21
SUMMER SIGNS	♋	Cancer	The Crab	June 21
	♌	Leo	The Lion	July 23
	♍	Virgo	The Virgin	August 23
AUTUMN SIGNS	♎	Libra	The Balance	September 23
	♏	Scorpius	The Scorpion	October 24
	♐	Sagittarius	The Archer	November 22
WINTER SIGNS	♑	Capricornus	The Goat	December 22
	♒	Aquarius	The Water-Bearer	January 20
	♓	Pisces	The Fishes	February 19

time, to which the difference in the longitude between the meridian on which the sundial is situated and the meridian on which the standard time zone is based must be applied to obtain *standard time,* which in Britain is GMT.

Depending on their construction, sundials determine the time by measuring the height of the sun *(altitude)* or by measuring the direction of the sun (bearing or *azimuth)* on a given date, or by a combination of both these factors. Accordingly, a sundial may be described as an altitude dial or as an azimuth dial, and all dials, other than altitude dials, are *directional* in the sense that they must be accurately aligned in order to function correctly.

Fixed sundials have been divided into two classes: primary dials and secondary dials. *Primary* dials are those drawn on the plane of the horizon, called *horizontal* dials, or those drawn perpendicular to it, on the planes either of the prime vertical (to face directly north or south) or the meridian (to face directly east or west), called *vertical* dials. To these are also usually added those drawn on the equinoctial and polar planes. Of these, the *equinoctial* sundial is the fundamental dial in the whole science of gnomonics from which all other dials may be derived. *Secondary* dials are all those which are drawn on the planes of other circles, besides the horizon, prime vertical, meridian, equinoctial and polar circles, namely those circles which either decline, incline, procline, recline or deincline.

Fixed sundials are traditionally adorned with a motto or saying, often with a reference to the passage of time. Such mottoes have been gathered together and published from time to time and may be found in various reference works.

One of the earliest so-called 'scientific' sundials showing the equal hours system, which we use today. The gnomon is missing but would probably have been a triangular piece of metal, jutting out at right angles to the wall. The sloping edge of the plate would have been aligned with the earth's axis, pointing to the pole, and its shadow would have indicated the time.

HISTORY AND DEVELOPMENT

The origin of the sundial is lost in antiquity, but the relationship between time and his own shadow would not have gone unnoticed by primitive man. So perhaps man himself was the first sundial. Likewise the direction and length of the shadows of different objects, trees, rocks or buildings would have been associated with the passage of time, both the time of day and the time of year. The latter was particularly important, just as it is today, for man needed to know the seasons of the year accurately, in order to know when to sow, when to gather in the harvest, when to prepare for the onset of winter, and when to expect storms and floods. A wooden staff or pole driven into the ground in a vertical position would have been the simplest form of indicator or gnomon which would have enabled man to determine such information from the direction and length of its shadow. The earliest recorded reference to a sundial goes back to the year 1300 BC in Egypt, whilst the earliest known Graeco-Roman sundial has been dated at 300 BC.

Most people think of a sundial as being the common or garden horizontal sundial and see it simply as a garden ornament rather than as an instrument for determining the time. However, for centuries in Britain vertical wall sundials were probably more common than any other form of dial and served both to regulate public clocks and to indicate the time to the passer-by. Sundials take many forms, shapes and sizes, but the vertical dial appears to have been the first fixed basic dial to have been in use in Britain. The earliest known dial of this kind can be seen on the Bewcastle Cross, in a remote churchyard in Cumbria, north of Hadrian's Wall and close by the remains of a Roman fort. Only the shaft of the cross remains, but it represents one of the finest surviving examples of Anglo-Saxon sculpture. It is elaborately decorated with

The Bewcastle Cross, Cumbria, is one of the finest works of late seventh-century Anglo-Saxon art. The earliest known sundial to be made in Britain is carved about two-thirds of the way up the shaft (close-up view above). The style hole, from which the hour-lines radiate, would once have had a wooden or metal peg inserted into it, to project at right-angles to the dial as an indicator or gnomon. The dial shows both the twenty-four hour time system and the old octaval tide system used to determine church services. The ornamental vine scrolls on the shaft show Mediterranean influence.

carved panels and shows a strong Mediterranean influence. The date of the cross is uncertain, but, from the evidence of the extensive runic inscriptions and from the quality of the sculpture, which suggests that it was a product of the golden age of Anglo-Saxon art, it probably dates from the late seventh century, about AD 685. The sundial itself also shows signs of Mediterranean influence. Indeed, the knowledge and skills required to construct the dial must almost certainly have had Near Eastern or Graeco-Roman scientific cultural origins. The sundial takes the form of a semicircle, with a style-hole at the centre, from which a series of downward radiating lines extends to the circumference. Originally a style or gnomon, in the form of a wooden or metal peg, would have been fixed in the style-hole, projecting at right angles to the vertical dial face towards the south. In sunlight, the shadow of the gnomon passing over the radiating lines would mark the time from sunrise to sunset. The lines divide the dial into twelve parts or hours of daylight, evidently in a system where there were twenty-four divisions counted in one day, measured from sunrise to sunrise.

It has been supposed that the Anglo-Saxons also used an octaval or tide system of time measurement, whereby the twenty-four hour day was divided into eight parts or *tides,* each of three hours duration. The word 'tide' is derived from the Anglo-Saxon term for time or hour. The terms 'morningtide', 'noontide' and 'eventide' are still occasionally used. However, the Christian church kept a strict system of its own to mark the regular times of prayer, according to the hours of the offices arising from the Passion of Christ. These church offices were at three-hour intervals, which were known as *canonical hours.* Of special significance were the third, sixth and ninth hours and it is clear that many Anglo-Saxon sundials were constructed primarily for the purpose of indicating these particular times or tides, since the corresponding hour-lines are usually marked with a cross at their extremity. In addition to the duodecimal (twelve division) hour system, the dial on the Bewcastle Cross is also marked in this manner for these three church offices. Thus, the Bewcastle dial is not only the first known prime-vertical sundial in Britain, constructed for the technical determination

A Graeco-Roman vertical sundial of the kind from which the earliest British dials may have been derived. It is situated at the corner of the south transept of the church at Orchomenos, Boeotia, in Greece, and may be of a later date than the Bewcastle dial. It is numbered from one to ten in Greek alphabetical numerals.

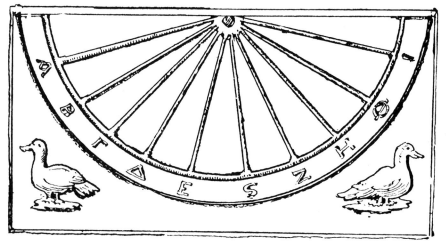

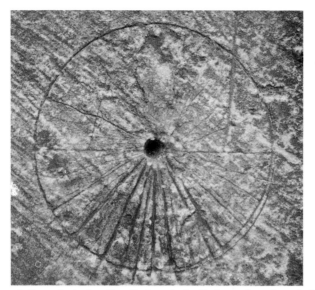

of time, but also the first ecclesiastical dial for the measurement of church offices.

Anglo-Saxon sundials should not be confused with those dials more commonly found scratched on the walls of churches, and known as *scratch dials* or *mass dials*. These are usually crudely scored into the stonework and are generally of a later date. Several such dials are often found close together on the same wall and it is supposed that they were the work of individual clerics, scoring the wall so that when the church was limewashed, as was once a practice, a small sundial could be painted and repainted readily, to mark the times of the church services. The Anglo-Saxon sundials, however, as well as being technical instruments, were, in many cases, finely

The vertical Anglo-Saxon sundial at Kirkdale Priory, North Yorkshire, dating from about 1064. The dial shows the octaval tide system of time measurement. The most important church offices are marked by a small cross at the extremity of the hour-line.

8

produced works of art. By comparison with the number of recorded scratch dials that have survived over the years (about two thousand), Anglo-Saxon dials are rare, but a number of beautiful examples can still be seen and appreciated. Bishopstone, in East Sussex, has a fine but simple Anglo-Saxon dial, probably dating from around AD 950, which, like Bewcastle, is marked for both the duodecimal system of time measurement and for the primary ecclesiastical offices. However, Kirkdale Priory in North Yorkshire has what is regarded as the best example of an Anglo-Saxon dial which has come down to us. It is considered to date from about 1064 and is in an excellent state of preservation, because a porch was built on to the building at a later date and has reduced the effects of wind and weather.

These Anglo-Saxon vertical sundials and the early medieval mass dials or scratch dials were far from being accurate. The primary cause of this was the use of a horizontal gnomon, which would have been perfectly satisfactory for a dial at or near the equator, but which would not give accurate readings in the higher latitudes of northern Europe or Britain. Only with the introduction of what might be described as the scientific sundial, with a gnomon inclined parallel to the axis of the earth, and which showed hours of equal length throughout the year, did the determination of time become a functional science. Whilst the technical benefits of inclining the gnomon parallel to the earth's axis may have been realised as early as the first century AD, it was not until the tenth century that the theory of equal hours was developed on a practical basis by the Arabs, who are thought to have devised the prototype of the 'modern' sundial. How and when this fundamental advance in the science of gnomonics was transmitted to Britain is unknown. It might have been during the Crusades, around the twelfth century, when northern European nations were brought into direct contact with Arabian science and culture, or through Moorish influence in Spain.

In the thirteenth century, mechanical clocks which sounded and showed twenty-four equal hours were coming into use. Whilst such clocks were complementary

The vertical Anglo-Saxon dial at North Stoke church, Oxfordshire. Note the iron gnomon projecting from the dial and the crosses marking the church services.

to the sundial, showing hours of equal length, they were unreliable and needed to be frequently checked. It was the sundial that served this purpose. By the fifteenth century, with the coming of the Renaissance, the division of the day into twenty-four equal hours had come into common use. Clocks and sundials flourished side by side, the one giving the time in cloudy weather and during hours of darkness, the other providing the means to regulate the clock, by determining the time of day from the position of the sun, during periods of sunshine. Over the next four hundred years, both clocks and sundials were produced in great profusion. Sundials, both portable and fixed, appeared in much variety and complexity, not out of necessity, but for artistic pleasure and to prove the mathematical skills and ingenuity of the diallist.

During the Renaissance the science of gnomonics, or the art of dialling as it was more generally known in Britain, became established as an important mathematical subject, closely related to astronomy and navigation, and its understanding was an accepted academic accomplishment in colleges and universities. Particular impetus was given to this dynamic science by the advent of the movable-type printing press in the latter half of the fifteenth

HOROLOGIOGRAPHIA.

The Art of Dialling.

teaching an eafie and perfect way
to make all kinds of Dials upon any
plaine plat howfoeuer placed:

VVith the drawing of the twelue Signes, and
houres vnequall in them all.

Whereunto is annexed the making and vfe of other Dials and Inftru-
ments, whereby the houre of the day and night is knowne:

Of fpeciall vfe and delight not onely for Students of the Arts Mathema-
ticall,but alfo for diuers Artificers,Architects,Surueyours of
buildings,free-Mafons,Saylors, and others.

By T. Fale.

AT LONDON
Printed by *Thomas Orwin*,dwelling in Pater nofter-Row out
againft the figne of the Checker. 1 5 9 3.

Title page of the first book in English to be devoted to sundials, written by Thomas Fale and published in London in 1593.

century, enabling the knowledge of this subject to be readily communicated. The earliest printed treatises on gnomonics appeared in Germany, France and Italy. The first work in English devoted to sundials was published in London in 1593 under the title *Horologiographia: The Art of Dialling* by Thomas Fale, a Cambridge mathematician. Thereafter, the publication of English dialling works grew steadily, just as the art of dialling itself progressed from the sixteenth century and flourished throughout the seventeenth and eighteenth centuries. Only with the greatly increased accuracy of mechanical timekeepers and the development of railways in the nineteenth cen-

tury did the usefulness of the sundial decline. Even at the beginning of the twentieth century a number of mechanical sundials were produced for use in areas where no time checks were readily available. Indeed, an advertisement for a sundial, intended as a functional instrument, for the serious purpose of determining time in remote places, was advertised as late as 1934 in the *Nautical Almanac*. Modern communications have now placed the art of dialling, as a scientific subject, into the history books. Nevertheless, this ancient art is part of our heritage and its study can give much pleasure.

A modern vertical direct east dial on the church of St Margaret of Antioch, Westminster. Note that the gnomon lies parallel to the dial plate, both being parallel to the polar axis. The dial is one of four which overlook Parliament Square and Westminster Abbey. They are painted blue, with the numerals gilded in platinum, and were designed by the author.

VERTICAL DIALS

Since vertical sundials, set up on walls in public places to regulate clocks and to give the passer-by the time of day, were the most easily seen, they were also the most common dials in general use. Some were quite plain, whilst others were extremely ornate. They were made of stone, slate or wood and were then fixed to the wall, or in many cases they were simply delineated directly on to the wall itself and painted. A vertical dial, set on a wall to face due south (that is, being delineated on the plane of the *prime vertical* circle), would be classed as a primary sundial and termed a *south* dial, or a *direct south* dial, or more particularly an *erect direct south* dial. The calculations for such a dial are simple and straightforward, as are the calculations for the other vertical primary dials, which face the cardinal points of the compass, namely a *direct north* dial, a *direct east* dial and a *direct west* dial. Nowadays, dials are often referred to as 'direct south-facing', 'direct east-facing' and so on, but the older dialling terminology was the language of dial makers in the centuries when dials were common everyday instruments and is more properly used.

Walls seldom face directly towards a cardinal compass point and consequently most vertical sundials are not direct but *decline* or face away from the compass point by so many degrees of arc. This angle of *declination* is always measured from due north or due south away towards the east or west. Sundials which decline in this way are classed as secondary dials, and they are more complicated to calculate and to delineate than primary dials. A dial which declines from the south towards the east is termed a *south-east vertical declining dial*, or a *south-east decliner*. Likewise, a dial declining from

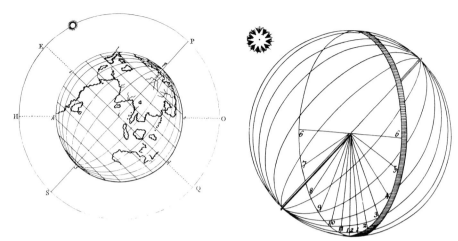

ABOVE LEFT: *The principles of dialling : the vertical direct south dial. The illustration shows the theory for the construction of a prime vertical sundial for a person in Scotland. One must imagine the earth as the terrestrial sphere, at the centre of the heavens, that is the celestial sphere. One must further imagine that the terrestrial grid system of meridians and parallels of latitude can be projected on to or have their exact counterparts on the celestial sphere, except that the one system rotates within the other, about the earth's polar axis. To an observer viewing this from the centre of the earth, however, if this were possible, it would seem that the celestial system rotated about the earth. In these systems: l is the person or place in Scotland; l - o is the latitude of the place; H - O is the observer's horizon, rationalised to become a great circle on a plane passing through the centre of the earth (and the centre of the celestial sphere); e - q is the plane of the equator; E - Q is the plane of the equinoctial; P - S is the polar axis. In this diagram, both systems have been stopped at a point where the sun is on the observer's meridian: in other words it is noon.*

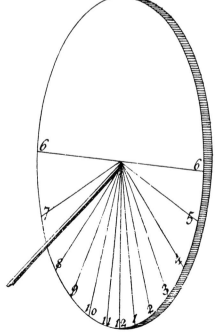

ABOVE RIGHT: *The principles of dialling : the vertical direct south dial. Imagine the earth being infinitely small at the centre of the celestial sphere and imagine a vertical dial of infinite size, delineated on the plane of that great circle called the prime vertical, extending out to the celestial sphere. The sun is on the sundial's 12 o'clock meridian (it is midday for the person in Scotland and he has the sun due south of him). The gnomon of the dial is aligned and coincides exactly with the polar axis. The meridian circles are all great circles whose planes cut the polar axis and the poles at 15 degree intervals. By projecting the planes of the meridians on to the plane of the dial plate, the hours can be marked off.*

RIGHT: *The principles of dialling : the vertical direct south dial. Imagine the celestial sphere removed and the dial to be of normal size. This is the vertical direct south dial, in its simplest form, which will show equal hours in the time-measuring system which is universally used today, delineated for the latitude of the person in Scotland. The theory for all other dials is based on these principles.*

12

LEFT: *A vertical direct south dial, dated 1790, at Clare, Suffolk. The gnomon is fixed directly on the twelve o'clock hour-line. The shadow shows the time to be exactly 2 p.m. (local apparent solar time).*

RIGHT: *A direct vertical north dial, engraved in stone and set into the wall of a church at Leighton Buzzard. Note that the gnomon, being parallel to the polar axis of the earth, points upwards on a north dial, not downwards as on a south dial. This same church has four other fine dials : one east, one west and two south.*

the north towards the west is termed a *north-west decliner*, and so on. A dial with a very large declination, that is, one which faces almost east or west, is termed a *great decliner*. Sometimes, perhaps for the pleasure of making the dial or simply out of vanity, the diallist would construct a double-declining dial, which consists of two declining dials, where one direct dial would have been sufficient. An example of this may be seen on St Peter's church at Evercreech, Somerset, probably put up in the eighteenth century. More rarely, one may come across a dial which combines perhaps three or four declining dials, on a square stone block, perhaps mounted above a porch but of less practical use and certainly constructed as a 'conceit'.

SUNDIAL FURNITURE

Sundials, particularly vertical sundials, frequently indicate much more than just the hour of the day. They were often designed to give such information as the position of the sun in the zodiac (that is, the date), the azimuth of the sun, the altitude of the sun, the number of hours elapsed since sunrise (sometimes called *Babylonian hours*), the number of hours elapsed since sunset on the previous evening (sometimes called *Italian hours*) and much more besides. This additional information, with which a dial may be furnished, is known as the *furniture* of the dial. To impress people, if not to expound the mathematical complexities of dialling, dial makers would sometimes fill sundials with such furniture, often making them difficult to read. At the National Maritime Museum, on the buildings of the Old Royal Observatory, there are some modern vertical sundials which demonstrate the use of individual items of dialling furniture and which show that they are quite easy to understand.

Amongst the Old Royal Observatory dials there is one which shows only the twelve o'clock hour-line and which indicates the moment of noon, when the sun is on the meridian. This moment has always been of great importance, since it provides the one instant during each day when the sun is at a known reference point and hence provides the basis for the determination of time during the twenty-four hour solar day. To this end, the *noon-mark* (sundial) was sometimes used

13

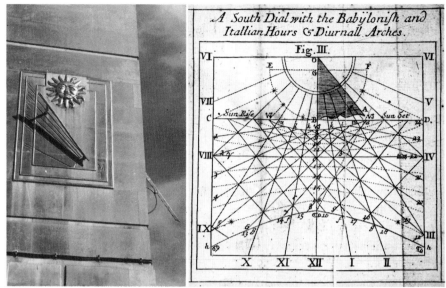

LEFT: *A vertical north-west 'great decliner' on the west-facing corner of a building in Cheapside, London. The term 'great decliner' applies because the wall on which the dial is situated declines some 75 degrees of arc from the north cardinal point of the compass towards the west. As a result of this large declination, the hour-lines are closely bunched together, whilst the dial centre, from which the hour-lines are delineated, is off the dial plate altogether. The dial shows local apparent time to be 3 p.m.*

RIGHT: *A direct south dial with its furniture (from 'Leybourn's Dialling Improv'd' by Henry Wilson, London, 1721). The dial illustrated shows the mathematical embellishments or furniture that could be applied to a basic sundial. Babylonian hours, so called, are the hours that have elapsed since sunrise, measured from the left-hand side of the horizontal line C - D, in a series of twenty-four hours from sunrise to sunrise. Italian hours, so called, are the hours that have elapsed since sunset on the previous evening, measured from the right-hand side of the horizontal line C - D, in a series of twenty-four hours from sunset to sunset. The diurnal arches are the lines which mark the sun's declination, or angular distance north or south of the equator. The movement of the gnomon's shadow or projected spot of light along these lines can indicate the date and the sign of the zodiac in which the sun is situated.*

to provide a daily time check, rather than by the use of the more conventional dial. Furthermore, the noon-mark dial could be furnished with the equation of time correction curve (or *analemma),* which enabled local mean time to be read directly off the dial. The Old Royal Observatory noon-dial is furnished in this manner and thus demonstrates the means by which clock time can be obtained. In this case, being at longitude zero, on the prime meridian of the world, the dial shows GMT. However, most sundials can be corrected for the difference in longitude (that is, the time difference between the standard time zone meridian and that of the sundial) and can be furnished with

equation of time curves for all hours. Such dials are rare, although fine modern examples, the work of the Reverend Father George Fenech, may be seen on a number of churches in Malta.

STAINED GLASS SUNDIALS

Of all the common vertical sundials, the least common and the most beautiful are the stained glass sundials which were built into the windows of mansions and churches in the seventeenth century and occasionally later. In these dials may be seen the mathematical skills of the diallist combined with the talents of the glass painter. The dials are calculated and delineated in exactly the same way as for

A vertical south-facing sundial showing hours from sunrise (horae ab ortu) in gold hour-lines, and hours from sunset (horae ab occasu) in black hour-lines. The spot of light indicates that 5 hours 20 minutes have elapsed since sunrise, and that 12 hours 40 minutes have elapsed since sunset on the previous evening.

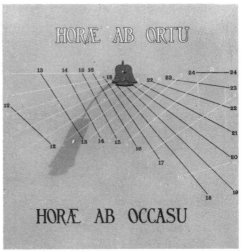

ordinary vertical dials, direct or declining, except that they are made of glass and built into a suitably situated window. The gnomon is fitted on the outside of the window, but the numerals are reversed and read from the inside, so that the viewer need not leave the comfort of the room on a sunny but cold winter's day.

Stained glass dials became popular in the latter part of the seventeenth century, when Puritan prejudice against colourful windows in churches obliged glass painters to look elsewhere for business, towards secular buildings, such as university colleges, civic halls, city mansions and the country houses of the landed gentry. One noted glass painter, Henry Gyles of York (1645-1709), apparently appealed to his friend Francis Place, the well known London engraver, who replied: 'I made Inquiry at Mr Price's about glass painters : he tells me there is 4 In Towne but not work enough to Imploy one, if he did nothing Else.' Nevertheless, as a result of this sad state

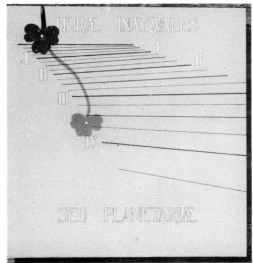

A vertical east-facing sundial showing unequal or so-called planetary hours (horae inaequales — seu planetariae). The spot of light indicates that the 'ancient' time, where the day from sunrise to sunset was divided into twelve equal parts or hours, is 3 hours 38 minutes near 21st June, giving local apparent time as about 8 hours 45 minutes (less 1 minute equation of time correction), GMT as 8 hours 44 minutes and BST as 9 hours 44 minutes, the actual clock time.

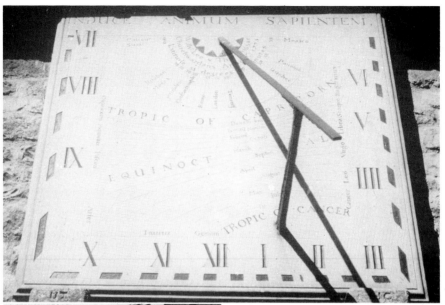

ABOVE: *A declining sundial on Eyam church, Derbyshire, dated 1775, with its furniture.*

LEFT: *A vertical noon-mark dial, showing only the 12 o'clock hour line, as illustrated in Bedos de Celles's work 'La Gnomonique Pratique', first published in Paris in 1760. The gentlemen are checking their watches, observing the moment when the spot of light, projected by the gnomon, is exactly on the noon-line.*

ABOVE: *A vertical direct south sundial, a noon-mark and mean-time dial (meridies media). When the spot of light, projected by the gnomon, crosses the vertical 12 o'clock line, the sun is on the meridian and it is 12 o'clock noon, apparent time. When the spot of light crosses the equation of time correction curve, for the corresponding date, as shown in the photograph, the dial will show 12 o'clock mean time, that is clock-time.*

ABOVE RIGHT: *A beautiful seventeenth-century direct south stained glass window sundial. Note the painted fly, which is very realistic, used by the dial maker as a pun on 'time flies' or as a common joke on his fellow men. The dial is at Lullingstone, Kent.*

RIGHT: *A vertical south-east declining sundial at Berkeley Castle, Gloucestershire. Note again the painted fly.*

LEFT: *The stumped cross pillar dial at East Hagbourne, Oxfordshire. The Puritan troops of Cromwell's army broke the cross in 1644, showing their disapproval of the established church.*

of affairs, the glass painter became more creative in his designs, turning from biblical scenes to heraldry, history, portraiture and so on, including glass sundials.

Since they are made of glass, there are not many of these dials in existence. Over the years, many have been broken or removed and they have usually disappeared without trace. The glass was thin and fragile, and it had to be drilled through in two or three places to allow the gnomon to be fixed in position. Consequently, it is fortunate that there are any examples at all of these glass sundials left.

PILLAR SUNDIALS

During the reign of Queen Elizabeth I in the sixteenth century, there was a Protestant backlash against the previous reign of the Catholic Mary, and some town and village preaching crosses were desecrated, for they were seen as symbols of the Roman Catholic religion. Likewise, during the Civil War, which began in 1642, many surviving crosses were reduced to stumps by the Parliamentarian troops. Again, they were seen as symbols of a wealthy church that supposedly supported the monarchy. The crosses had been set up originally to remind the traveller of his Christian beliefs and to encourage him to give prayers of thanksgiving for his safe arrival at each stage of his journey. They also served as meeting places, where clergy could preach to a congregation if there were no church, or where people could simply voice their opinions. Frequently they were set up in market places at the centre of the town and were referred to as market crosses.

Despite the turbulent times, the art of dialling flourished, and it was this art that provided a use for the stumped cross. By placing a stone block on which a dial or dials had been carved on top of the pillar where the cross itself had been, the structure was restored as an object of

LEFT: *The west face of the pillar dial set up in 1698 at Maud Heath's Causeway, Chippenham, Wiltshire.*

18

LEFT: *The market cross sundial at Witney, Oxfordshire, set up in 1685.*

RIGHT: *A classical eighteenth-century type of pillar dial, close to York Minster.*

dignity if not of religious significance. Thus the 'cross', particularly the market cross, continued to serve as a meeting place, with the added benefit of a sundial to tell the time of day. Many of these crosses were roofed over, to give shelter from the weather, but some new seventeenth-century market crosses were purposely built to support a sundial. Such a dial would usually take the form of a direct south sundial, or, if more than one dial was provided, they would be direct vertical dials, which were easily placed to face the cardinal points of the compass. Oakham, Witney and Woodstock, for example, each had a market cross of this kind, although the only trace of the Woodstock cross is a stone remnant, supposedly the actual dial, built into the wall of the town hall.

With the Restoration of the Monarchy in 1660, the art of dialling continued to flourish and increase in popularity. Elegant dials were made and mounted in conspicuous places to indicate the time.

In the same spirit, pillar dials were built with classical grace as primary features of villages, towns and cities. Although many such dials were pulled down long ago, notably the famous Seven Dials and Covent Garden dials of London, there are still a number of fine examples to be seen throughout Britain.

At East Hagbourne near Didcot in Oxfordshire there is a typical example of a seventeenth-century stumped-cross dial, whilst York has a 'classical' pillar dial of the late seventeenth or early eighteenth century (albeit with the gnomon of the north dial fixed in the wrong way). Other remarkable detached pillar dials include the one known as the Countess's Pillar, a massive column set up in 1656 by Anne Clifford, Countess of Pembroke, at a lonely spot on the wayside between Brougham and Appleby in Cumbria, and the roadside dial at Chippenham causeway, Wiltshire, set up in 1698.

The forward-inclining or proclining component dials of a Scottish multiple sundial. These component sundials are also deinclining dials and so are sometimes termed declining dials. Note the remains of the metal gnomons, which have rusted away. Such dials may well have never caught the rays of the sun but were extravagant mathematical 'conceits' of the diallist's art.

MULTIPLE DIALS

Whilst the basic utilitarian dial for public purposes was the common vertical sundial, the art of dialling bloomed in sixteenth- and seventeenth-century Britain in many complex mathematical forms, which were particularly manifested in *multiple* sundials. These were usually but not always *detached* dials, as were pillar dials, that is, they were not attached to a building. The evidence suggests that the earliest English multiple sundials may have had their origins in Germany and France. A German scholar and astronomer, Nicolaus Kratzer, came to England, perhaps in early 1518, to the court of Henry VIII to be appointed 'deviser of the King's *horologes*' (sundials). Furthermore, there are interesting similarities between the dials illustrated in the works of Sebastian Munster, an eminent gnomonical authority, and the earliest known multiple dials in England,

which are probably those at Elmley Castle, possibly dating from about 1545. Nicolaus Kratzer set up an elaborate pillar multiple dial outside St Mary's church, Oxford, in about 1523, and another in the garden of Corpus Christi College nearby. Although almost all colleges had their own sundials, few sixteenth-century examples survive and Kratzer's dials had disappeared by the early eighteenth century. However, the restored Turnbull Dial, often called the Pelican Dial, which stands in the front quadrangle of Corpus Christi, is a beautiful example of an early multiple dial although it is not the original dial of 1579.

A multiple dial, as the term implies, is a sundial which comprises numerous individual dials as component parts of the sundial as a whole. For the most part, multiple dials were intended for decorative purposes to show off the mathematic-

LEFT: *A multiple sundial in the churchyard at Elmley Castle, Hereford and Worcester. Whilst much of it has been restored, it is perhaps the earliest dial of its kind in England, thought to date from about 1545. Note the hollowed scaphe dials.*

RIGHT: *One example of many splendid Scottish multiple sundials.*

al knowledge and skills of the diallist. On examining such a sundial, one may find a whole range of small component dials, including almost every class of dial described — horizontal, equinoctial, polar, vertical (direct and declining) as well as inclining/proclining, reclining and deinclining dials — together with a variety of cup and scaphe dials, hollowed out of the stone in different shapes and sizes.

Whilst the earliest multiple dials to be found in England may exhibit German or French influence, the most extraordinary multiple dials are found in Scotland and their origin is a matter for speculation. Undoubtedly the Scottish mind for mathematics and science took to the multiple sundial in a way which is found nowhere else in the science of gnomonics or the art of dialling. These magnificent multiple dials are unique to Scotland and show a remarkable passion for dialling in all its most complex mathematical forms. These dials have been subdivided into classes: obelisk-shaped, lectern-shaped

and facet-headed dials. The typical *obelisk* dial comprises a square stone shaft, above which is placed a bulged octagonal-shaped capital, with a tapering finial above, the whole reaching a height of about 7½ feet (2.3 m). On every surface are incised all manner of dials, some with gnomons, others without but using the edge of a piece of stonework to cast the shadow. *Lectern* dials, as the term implies, resemble a lectern but again contain dials of all shapes and sizes. The *facet-headed* dials also contain large numbers of small dials, but, as the name implies, the bulk of these are combinations of vertical declining and inclining dials. Such an edifice sometimes resembles the ornate head of a mace, once carried into battle by knights in armour, and may contain as many as two hundred individual dials or more. The dials themselves are of no great accuracy, but they do illustrate every conceivable form of traditional dial.

21

ABOVE: *An eighteenth-century polyhedral sundial at Penshurst Place, Kent. It illustrates the alignment of the gnomons with the polar axis. It also illustrates a polar sundial, above the vertical direct south dial, and a direct proclining dial immediately below.*

RIGHT: *The famous column in Corpus Christi College, Oxford, known as the Turnbull dial. It is a multiple dial, originally set up in the sixteenth century.*

A beautifully engraved seventeenth-century double horizontal sundial, by Thomas Tuttell. The leading edge of the shadow cast by the gnomon indicates the local apparent solar time to be a few minutes before 2 p.m. The vertical support to the gnomon also casts a shadow, the edge of which, if followed to the point where it intersects the hour-line on the planispheric projection of the celestial sphere, corresponding to the time indicated by the dial itself, will give the position of the sun in terms of date, declination, altitude and azimuth. Thus the dial serves a dual purpose, hence the term double horizontal dial. It was invented by the English mathematician William Oughtred and published in his work 'Description and Use of the Double Horizontal Dial' (London, 1636).

HORIZONTAL DIALS

The common horizontal garden sundial is classed as a primary dial and is a dial described on a horizontal plane or a plane parallel to the horizon. It is normally a detached dial, set upon a plinth away from a building in a situation where it will receive the maximum amount of sunshine all the year round.

Horizontal sundials became fashionable in the sixteenth century in England, both as garden ornaments and as instruments for marking the passage of time. The earliest sundial of this kind in England is uncertain, although there is in existence a horizontal dial with an apparent date of 1395 crudely cut into the underside of the dial plate, on which an earlier dial had been engraved, but had seemingly never been used. The style of

the engraving appears to be of a later period than the fourteenth century, more like that of the sixteenth century, and there seems to have been no good reason why this earlier dial should not have been made use of, unless an error was made in cutting the date into the metalwork, so obliging the dial maker to start afresh on the reverse of the plate.

The National Maritime Museum has in its sundial collection, in the Old Royal Observatory, a fine square bronze dial dated 1582, signed by the noted instrument maker Humphrey Cole. A number of other examples of sixteenth-century horizontal sundials, if not still in use, are to be found in the scientific instrument collections of various museums. Horizontal sundials have remained popular as

ABOVE LEFT: *A fine early eighteenth-century common or garden horizontal sundial at Lacock Abbey, Wiltshire. The gnomon, being aligned with the polar axis, is therefore also angled to the value of the latitude of the place for which it is made.*

ABOVE RIGHT: *A twentieth-century horizontal dial at Shakespeare's birthplace at Stratford-upon-Avon.*

BELOW: *An engraving, in the collections of the National Maritime Museum at Greenwich, showing a plan of the double horizontal sundial by Thomas Tuttell, 'Mathematical Instrument maker to the King's most excellent Majesty at Charing Cross, London'. The planispheric projection takes the form of a grid system. The sun's altitude is obtained by reference to the rule shown to the right on the dial and extending from the index of the dial support, or secondary gnomon. Note that the equinoctial is shown on the projection, as well as the tropics of Cancer and Capricorn.*

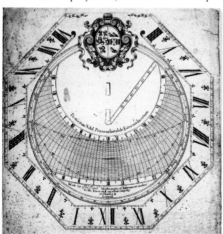

garden ornaments to this day, being simple to read, attractive to look at, and requiring little maintenance. However, during the seventeenth and eighteenth centuries, like their vertical counterparts, well made horizontal dials were often engraved with useful furniture, including devices to give the position of the sun in the zodiac, its bearing and altitude at any moment, and the values of the equation of time. Most good instrument makers also signed and sometimes dated their dials, which were used by their owners for the serious purpose of checking their household clocks and watches, as well as serving as centrepieces for the garden. Many of these fine dials can still be seen in the grounds of large country houses.

A modern basic equinoctial sundial, by Wendy Taylor, at the entrance to St Katharine's Dock, London. Technically it should be termed an 'upper' equinoctial dial and will show the time only while the sun is north of the equator.

EQUINOCTIAL DIALS

The most important primary class of sundial, and the fundamental dial from which all others are derived, is that known as the equinoctial dial, that is, one described on an equinoctial plane, or a plane representing that of the equinoctial. The equinoctial is the plane of the equator extended to the celestial sphere (the imagined sphere of the heavens), that is, the celestial equator. It is so called because the *ecliptic* or apparent path of the sun intersects this great celestial circle at two points, and because the length of the day equals the length of the night (whence the term *equinox)* when the sun reaches one or other of these points. Thus an equinoctial dial, in its most basic form, is a flat plate or ring set to lie parallel to the plane of the equator, with a gnomon passing through its centre at right angles to the plane of the dial. Sometimes such a dial may be termed an *equatorial* dial, which is not correct in original dialling terminology but which has come to be accepted

through common usage. Furthermore, the basic concept of the equinoctial dial has become extended to include those dials which have an equatorial hour-ring, on which the time is indicated by the gnomon. In this category, the *armillary* sundial is the one kind of dial that illustrates the whole principle of the science of gnomonics. It consists of a number of metal rings representing the circles of the celestial sphere, usually with a broad hour-ring representing the equinoctial or celestial equator on which the numerals of each hour of the day are engraved or painted. A metal rod passes through the centre of this assembly of rings, parallel to the polar axis and at right angles to the hour-ring, and its shadow indicates the time. Sometimes a small metal ball is fitted at the centre of the rod, representing the earth. When the shadow of this ball passes along the centre line of the hour-ring during the course of the day, provided that the dial is accurate, the sun will be at one or other

25

LEFT: *An armillary dial, showing the equinoctial hour-ring. The gnomon is the polar axis of this representation of the celestial sphere.*

BELOW LEFT: *A globe sundial, made of stone. The time is indicated by rotating the metal gnomon until the point is directed at the sun, when there will be no shadow. One then reads off the hour which is directly under the gnomon.*

BELOW RIGHT: *A modern hemispherical equinoctial sundial, designed by G. P. Woodford.*

LEFT: *A mechanical equinoctial sundial, invented by Gibbs and first produced in 1902. It is of the type termed a 'heliochronometer' and compensates for the equation of time automatically.*

RIGHT: *A Gibbs heliochronometer of the early twentieth century. By setting the month disc against the day scale, a cam adjusts the upper perpendicular pinhole vane (the gnomon) for the value of the equation of time. The dial plate is then turned until the sun is directly in line with the two vanes, causing a spot of light to be projected on to the centre line of the lower vane. Mean time is then read off the scale fixed to the rim of the dial.*

of the equinoxes.

The equinoctial sundial is not only the dial least likely to be in error in its construction, but it is also the dial which lends itself most to the pursuit of accuracy, so far as accuracy may be achieved by a sundial. As clocks improved in performance, so instrument makers sought to improve the accuracy of the sundial. In the latter part of the seventeenth century the Reverend John Flamsteed, the first Astronomer Royal at Greenwich, produced an accurate table for the values of the equation of time — the difference between apparent solar time (sundial time) and mean solar time (clock time). William Molyneux printed Flamsteed's equation of time table in his book *Sciothericum Telescopicum,* published in Dublin in 1686, in which he sought to make improvements to the sundial. Early in the eighteenth century, the first dials to include this correction appeared, and the equation of time *analemma,* or figure-of-eight correction curve, took an important place in dialling literature. Nevertheless, it was not until the nineteenth century, with the advent of the railways, bringing rapid communication between cities and towns and creating an even greater need for accuracy in clocks and watches, that any marked improvements were made to the sundial. However, near the end of the nineteenth century, mean-time sundials, allowing for the equation of time correction, and showing clock time, were coming into use. Furthermore, mechanical equinoctial mean-time dials, or *heliochronometers,* were taking their place as useful accurate instruments. In France heliochronometers were used well into the twentieth century to check the times of departure of express trains. Whilst the coming of the electronic age has eclipsed the sundial as a scientific instrument, modern mean-time dials are still to be seen and enjoyed. A modern example of this kind can be seen at the National Maritime Museum, commissioned to celebrate Her Majesty the Queen's Silver Jubilee in 1977. It is a fine sculpture in the form of two dolphins, whose tails cast a shadow on to the graduated dial plate. The gap between the shadows of the tails indicates the time to within a minute.

A memorial cross dial. It is aligned towards the sun, not away from it, and lies in the plane of the equator/equinoctial. It combines both polar dials with direct east and west dials. The hour scales are engraved on the sides of the cross.

POLAR AND OTHER UNUSUAL DIALS

The last primary class of sundial is the so-called *polar* dial, which is one described on a plane passing through the poles of the celestial sphere and the east and west points of the horizon. It is an exceedingly rare dial, although it is partly manifest in the form of a memorial cross dial, which is a combination of direct east, west and polar dials. The gnomon, in the case of the polar part of the dial, is the uppermost east or west edge of the 'vertical' cross-piece, whilst the hour-lines are engraved on the uppermost surfaces of the 'horizontal' cross-piece.

Whilst all dials have their mathematical origin in the celestial sphere, there are some which stand out as being unusual or in a class of their own. Nevertheless, they can usually be placed without too much difficulty in one of the classes described.

The unusual multiple sundial on the buttress of Bleadon church, Avon. The component dials comprise various hollowed scaphes, in which the stone edges are designed to be the gnomons. The uppermost dial is a rare and interesting cup dial with a short metal gnomon, perpendicular to the base of the cup and parallel to the plane of the equinoctial.

28

ABOVE: *A simple but remarkable west-facing scaphe dial, cut into a tombstone in the churchyard at Saxmundham, Suffolk. There is a similar scaphe dial on the east face of the stone. The upper edge of the scaphe lies in the polar axis and acts as the gnomon.*

RIGHT: *An unusual equinoctial sundial, shaped rather like an anchor, set up in 1825 in the Deanery garden at Rochester, Kent. The flukes of the 'anchor' act as gnomons. On the shaft of the dial there is an engraved plate with a table giving the values of the equation of time.*

BELOW: *A modern multiple scaphe dial, set up as a memorial in a churchyard at St Neot, Cornwall. The dial was designed by Dr Grylls.*

LEFT: *A modern sculpted vertical south-west declining sundial on the wall of the Marine Society's headquarters at Lambeth, London. The dial combines the symbols of the Nautical Institute, an armillary dial, and the Marine Society, a sea-dog. The dial shows both standard time and summer time. It was designed and delineated by the author, sculpted by Edwin Russell and unveiled by Her Majesty the Queen in 1979.*

BELOW LEFT: *A modern sculpted scaphe dial, in the form of apple blossom, in which a hole allows the sun's rays to project a spot of light on to a 12 o'clock hour-line. The dial faces north.*

BELOW RIGHT: *A detail of the north-facing apple-blossom dial, showing the spot of light on the noon-line. There is a south-facing apple-blossom dial as well.*

A horizontal cannon sundial (early twentieth century). The dial plate is made of marble and the fittings are of brass. The gnomon indicates the normal local apparent solar time. However, by priming the cannon with the proper black powder and by setting the lens to the correct angle for the noonday altitude of the sun, at the right moment, when the sun arrives on the meridian, the concentrated rays of the sun will ignite the powder and cause the cannon to fire with a suitable bang. In bygone days, this provided a primitive time signal.

FURTHER READING

A number of popular late nineteenth-century and early twentieth-century books and articles on sundials include a page or two on how to construct a sundial. To keep the instructions simple, the authors seldom explained the mathematical principles on which the art of dialling is based. There are, however, a number of good books concerned solely with the construction of sundials, some of which are included in the following list.

Daniel, Christopher St J. H. *Sundials on Walls*. Maritime Monographs and Reports number 28, National Maritime Museum, Greenwich, 1978.
Drinkwater, Peter I. *The Art of Sundial Construction*. Drinkwater, Shipston-on-Stour, 1985.
Green, Arthur R. *Sundials: Incised Dials or Mass Clocks*. SPCK, London, 1978.
Mayall, R., and Mayall, M. *Sundials: How to Know, Use and Make Them*. Sky Publishing Corporation, Cambridge, Massachusetts, USA, 1973.
Pattenden, Philip. *Sundials at an Oxford College*. Roman Books, Oxford, 1979.
Waugh, Albert E. *Sundials: Their Theory and Construction*. Dover Publications, New York and London, 1973.

SPECIALIST SUNDIAL BOOKSELLERS
Rogers Turner Books Ltd, 22 Nelson Road, London SE10, and 24 Rue du Buisson Richard, 78600 Le Mesnil-le-Roi, France. Telephone (London): 01-853 5271.
Rita Shenton, 148 Percy Road, Twickenham, Middlesex TW2 6JG. Telephone: 01-894 6888.

SUNDIAL MAKERS

Brookbrae Ltd, 53 St Leonard's Road, London SW14 7NQ. Telephone: 01-876 4370.
South West Sundials, Sundial House, 15 Chesterfield Road, Laira, Plymouth, Devon PL3 6BD. Telephone: Plymouth (0752) 27582.
Sundials, 5 Elm Grove, Taunton, Somerset TA1 1EG. Telephone: Taunton (0823) 72400 or 74681.
Wall Sundials by Post, Frank Evans, 15 Thirlmere Avenue, Tynemouth, Tyne and Wear NE30 3UQ. Telephone (evenings): Newcastle-upon-Tyne (0623) 575354.
G. P. Woodford, 26 Rue Lamartine, 29120 Pont L'Abbé, France.

PLACES TO VISIT

Sundials can be seen in many places. The list of places to visit consists mainly of places where more than one sundial can be seen and of museums where there are sundial collections. The gardens of many other historic houses and National Trust properties contain interesting dials. Intending visitors are advised to find out the times of opening before making a special journey.

Drummond Castle Gardens, Muthill, Tayside.

Hever Castle, Hever, near Edenbridge, Kent. Telephone: Edenbridge (0732) 865224.

Museum of the History of Science, Old Ashmolean Building, Broad Street, Oxford OX1 3AZ. Telephone: Oxford (0865) 243997.

National Maritime Museum (including the Old Royal Observatory), Romney Road, Greenwich, London SE10 9NF. Telephone: 01-858 4422.

Penshurst Place, Penshurst, near Tunbridge Wells, Kent. Telephone: Penshurst (0892) 870307.

Royal Museums of Scotland, Chambers Street, Edinburgh EH1 1JF. Telephone: 031-225 7534.

Science Museum, Exhibition Road, South Kensington, London SW7 2DD. Telephone: 01-589 3456.

Whipple Museum of the History of Science, Free School Lane, Cambridge CB2 3RH. Telephone: Cambridge (0223) 358381, extension 340.

A modern analemmatic sundial, designed and delineated by the author for the Liverpool International Garden Festival in 1984. In principle it is an azimuth dial, with the time being indicated by a vertical gnomon, in this case a person, standing on a scale of dates on the central stone paving slab. The surrounding stone paving slabs, marking the hours, all lie as points on the path of an ellipse. This dial is laid out to give summer time. The photograph shows that the shadow is indicating the time to be 12 o'clock (BST).